Understanding
Just in Time

Malcolm Wheatley

© Copyright 1997 by Barron's Educational Series, Inc.

First published 1992 in Great Britain by Hodder and
Stoughton Educational, copyright © 1992 Malcolm Wheatley

All inquiries should be addressed to:
Barron's Educational Series, Inc.
250 Wireless Boulevard
Hauppauge, New York 11788

Library of Congress Catalog Card No. 96-49056

International Standard Book No. 0-7641-0126-9

Library of Congress Cataloging-in-Publication Data
Wheatley, Malcolm.
    Understanding just in time / Malcolm Wheatley,
        p.   cm. — (Barron's business success series)
    Includes index.
    ISBN 0-7641-0126-9
    1. Industrial productivity.   2. Just-in-time systems.
    I. Title. II. Series
    HD56.W49 1997
    658.5'6—dc21   96-49056
                                                    CIP

PRINTED IN HONG KONG
987654321

Contents

◆

Chapter 1

Introduction

Companies all over the world are changing the way that they manufacture their products. This in itself, of course, is nothing new. Businesses have always fine-tuned their production processes in order to shave a little more off production costs or to become slightly more efficient. Some of these companies, however, are rather special.

Two things make these particular companies stand out. The first is the sheer scale of what is going on—not only the scale of the changes that they are making, but also the scale of the improvements that are being achieved. The second is that they are usually changing the way that they are doing things to the exact opposite of the way that they were doing them before. Where they used to make bigger batches of products, they now make smaller batches. Where they used to employ large numbers of inspectors and still suffer from poor quality, they now employ far fewer, but achieve levels of quality vastly higher than before. And where they used to manage to make only small improvements in efficiency, they now make enormous strides.

These companies are implementing Just in Time. This chapter explains how Just in Time evolved and how it fits in with some of the other approaches and manufacturing techniques that a company might have in place.

We look at these topics:

◆ how Just in Time evolved

◆ Just in Time vs. other approaches to manufacturing

HOW JUST IN TIME EVOLVED

Understanding where Just in Time came from and how it was originated helps in understanding what it is and how it works. Most people who have heard of Just in Time have also heard that it is a Japanese invention dating from the 1960s. Japan's position in the world marketplace then was not at all what it is now. At that time Western products were regarded by consumers as being far superior to Japanese ones, which were generally thought of as poor quality imitations.

Now, of course, in products ranging from photocopiers to cameras, stereo systems to cars, the complete opposite is true. Japanese products are highly regarded and consumers often prefer to buy Japanese brands rather than Western ones. One of the biggest factors contributing to this transformation was the Japanese development of Just in Time.

As a nation rebuilding its economy after the devastation of World War II, the Japanese recognized that they had few natural advantages. Basic raw materials usually had to be imported—perhaps from the very countries to which the finished goods were to be re-exported. Land prices and building costs were high, so factories and warehouses had to be small, which supposedly made them inefficient. And prices, product quality, and on-time delivery had to be far better than Western levels if Western customers were to be persuaded to deal with Japanese factories thousands of miles further away than their present suppliers.

In order to be competitive, therefore, the challenge facing Japanese factories and Japanese managers was truly remarkable: reliable deliveries of lower cost and higher quality products had to be squeezed from cramped, and sometimes antiquated, factories.

While a number of different factors helped the Japanese achieve this feat, three are usually regarded as being the principal ones:

◆ lack of space

◆ the adoption of new quality techniques

◆ the development of the Toyota Production System

LACK OF SPACE

In part, this was luck. Cramped factories are nothing new. Plenty of Western companies have moved from constricted city center locations to less built up areas in order to generate more space for themselves. Perhaps because this wasn't such an option, Japanese managers had to find a way of working within these tight space constraints, rather than simply moving away from them to bigger premises.

Think of a factory: how much space is given over to just storing material that is waiting to be used? How much is

◆ raw materials and components?

◆ work in progress between machines and production processes?

◆ finished goods waiting to be distributed or sold?

Japanese companies started to recognize that most of the space in a typical factory is simply used for storage. It is not being used for actual manufacturing.

Raw materials

Raw material storage space could be dispensed with by not holding any raw materials. So companies started buying only from suppliers who could deliver raw materials directly to the factory floor as they were required, rather than to a separate factory warehouse where they were first stored rather than being used. This is partly the origin of the term Just in Time.

Work in progress

Work in progress was a little more difficult to remove. But by using the techniques covered in Chapter 6, manufacturing batch sizes and lead times—the two levers that determine work in progress levels—were reduced and then reduced again. Then, by eradicating for good the storage space between machines and production stages or departments, the new lower levels became built into the process.

Finished goods

Finished goods storage space could also be eliminated. If a company was itself a supplier to a Just in Time customer, this tended to happen automatically. If the customer wanted his supplies in once-a-day deliveries, managers soon realized that it was easier (and usually more profitable) to manufacture the goods in once-a-day batches. For those who weren't Just in Time customers, the need to hold "just in case" buffer stocks diminished as well. With the faster manufacturing lead times that came with the lower levels of work in progress, customer orders could be met more quickly, avoiding the need to hold so much stock.

ADOPTION OF NEW QUALITY TECHNIQUES

The problem with getting rid of all the stocks of materials and work in progress was that it also removed a lot of the safety margin. If the parts that a supplier delivered were of poor quality and had to be rejected, production stopped. On the production line, manufacturing problems or scrapped batches often meant the risk of late deliveries to the customer. In turn, low stocks of finished goods meant that the traditional buffer between manufacturing problems and the customer—lots of stock on the shelves—wasn't there. Better quality was vital.

Defect prevention

Ironically, the solution adopted by the Japanese was to import an answer from the West. For many years Dr. W. Edwards Deming had been preaching—practically unheard—an approach to quality control that lay in defect *prevention* rather than defect *detection*.

This was just what the Japanese needed: a way of stopping problems from occurring, rather than in successfully finding them at final inspection. Dr. Deming was invited to visit Japan in 1950 to lecture on quality to Japanese companies. Today, many of these companies publicly acknowledge their debt to his lectures. The main lobby in Toyota's Tokyo headquarters has three photographs in it: one of Toyota's founder, one of Toyota's current chairman, and one of Dr. Deming.

We will cover Dr. Deming's approach to quality in detail in Chapter 4. The basic messages that he took to Japan, though, were simply put. Looking at some of them, it is easy to see how again they contrast with the approach adopted by traditional Western companies.

TRADITIONAL APPROACH

◆ Inspection is the key to quality.

◆ Quality control is a cost.

◆ Buy from lowest cost suppliers.

◆ Play suppliers against each other.

◆ Quality comes from quality control.

DEMING APPROACH

◆ Produce defect-free goods; eliminate inspection.

◆ Good quality is more profitable than poor quality.

◆ Buy from suppliers committed to quality.

◆ Work with suppliers.

◆ Quality comes from top management's commitment.

THE DEVELOPMENT OF THE TOYOTA PRODUCTION SYSTEM

In parallel with what the Japanese were learning from Dr. Deming about productivity, Toyota was developing an approach to production under the guidance of Mr. Taiichi Ohno, a production engineer working for the company.

Recognizing that Dr. Deming was advocating low levels of inventory as an aid to quality control, Toyota removed assembly line buffer stocks completely, allowing workers to stop the line if there was a problem—and everybody else would come and help sort it out.

For manufacturing operations that were not linked by an actual assembly line, Mr. Ohno developed a system known as *kanban* to act as an imaginary assembly line instead, linking operations together as if there were a real assembly line between them.

In Chapter 6, we will learn about the *kanban* system and how it builds an imaginary assembly line not only within factories but also between the customers of finished products and the suppliers of the raw materials that go into them.

Many of the other techniques we will be learning about—such as set-up time reduction—also originated at Toyota under Mr. Ohno. He is sometimes called the father of Just in Time.

JUST IN TIME AND HARLEY DAVIDSON'S TURNAROUND

In the 1970s and early 1980s Harley Davidson was an inefficient, low quality motorcycle manufacturer that had effectively almost destroyed their previous excellent reputation, was short of cash, and in 1982 was on the verge of bankruptcy. A new enlightened management team had taken over in 1981 and brought some of the latest American and Japanese management ideas to the company. They quickly introduced Just in Time parts inventory control to their manufacturing process and in 1982 generated enough cash to save the company from Chapter 11 bankruptcy. Inventory turns went from five to twenty, inventory levels came down 75 percent, and productivity increased by 50 percent. To achieve this, Harley realized that it had to introduce a continuous quality improvement program, because it is essential that if you are only going to have a few parts on the line, they all must be high quality. This program had three elements, Employee Involvement (EI), Just in Time inventory (JIT), and Statistical Operator Control (SOC).

Harley Davidson completely changed its management methods, its production systems, and its relationship to its employees. By inviting employees to become the engine of change, by enlightening them with the latest Japanese and American industry quality and efficiency ideas and methods, and by equipping them with the tools they needed to achieve this, management succeeded in turning the entire Harley Davidson operation around.

This company is now producing some of the highest quality motorcycles manufactured anywhere in the world, desired the world over, and in the process has become a very profitable company.

THE HONDA PRODUCTION SYSTEM IN THE UNITED STATES

In 1978, Honda of America Manufacturing Inc. was established and motorcycle production commenced in Marysville, Ohio. In 1979, automobile production commenced in an adjacent plant in Marysville. Honda brought their efficient manufacturing philosophy with them to these plants in the United States. This included a heavy emphasis on high quality motorcycles and automobiles, driven by what is now known as continuous improvement in the entire design and manufacturing process, and Just in Time inventory control.

These Honda plants carry parts sufficient for a few production hours (local suppliers) or a few days (supplies from other states or

Japan). Honda has energetically worked with U.S. suppliers to make them partners in Honda's manufacturing process, and in most cases has trained them how to keep their own inventories of parts down to a few hours worth of their own production need, thereby removing expensive inventory from the entire system. Honda has also helped and encouraged its suppliers with the quality improvement systems that are necessary and vital to achieve the level of quality that allows Just in Time to work.

JUST IN TIME VS. OTHER APPROACHES TO MANUFACTURING

Business these days is full of new acronyms or buzzwords, and manufacturing seems to be no exception. Some of the different approaches complement each other in that they can often work together to create a more powerful whole. Others, though, are direct opposites, and impede progress by aiming for conflicting goals.

The main approaches to manufacturing that are usually found in companies are

◆ Reorder Point Control (ROP)

◆ Materials Requirements Planning (MRP I)

◆ Manufacturing Resource Planning (MRP II)

◆ Optimized Production Technology (OPT)

◆ Just in Time (JIT)

It is important to understand the strengths and weaknesses of each approach, and how each of them fits in with Just in Time. Some have few conflicts with Just in Time; others don't. Companies can waste a lot of effort by trying to adopt two conflicting approaches in parallel!

ROP—REORDER POINT CONTROL

ROP is the way that companies have historically organized their manufacturing processes. A basic version of the idea is the two-bin system that is still in use today, where the stock on hand of each component or raw material is held in two separate bins. When the stock in one bin runs out, an order is placed for some more. By the time the second bin has run out, the replacement order has arrived, allowing both bins to be refilled. Often nowadays computers have automated the process, allowing the stock to be held in one physical location, rather than two. They also use statistical formulae to produce better estimates of demand and suitable safety stocks.

With ROP, all the way through the manufacturing process stocks of materials and parts are continually being used up and replenished. It is a complex exercise, and difficult to get right, as none of the stocks are linked together in any way. They are all independent, and so stock-outs are common as the order mix changes.

ROP doesn't really fit in with Just in Time. As we will learn later in the book, Just in Time is all about establishing linkages to synchronize flow to meet customer demands. Reorder point systems have very little flow at all.

MRP I—MATERIALS REQUIREMENTS PLANNING

The idea behind MRP I is to try and coordinate all these separate stockholdings. With MRP I, a top level demand for the finished products that a company makes—for example, washing machines—is exploded to yield the demands for the various individual components that go into it, such as motors, valves, and piping. These demands are then compared with the actual physical stocks of each component, as well as those due from suppliers, to specify the new orders for each part that have to be placed and when the deliveries have to arrive.

At first sight, MRP I is certainly a closer fit with Just in Time than is ROP. Where it falls down, though, is that it doesn't help managers to run factories at the level of their individual machines. It is a top-level approach. Typically, MRP I tells companies how many of a particular part to make each week or each day. It doesn't say precisely when they need to be made in order to reach the next machine or process just as it runs out of work to do, nor does it update its production plan minute by minute to reflect what is actually happening on the factory floor. Just in Time does—and it also helps improve quality and reduce costs.

MRP II—MANUFACTURING RESOURCE PLANNING

As companies implemented MRP I, they frequently found that because of capacity problems, they weren't capable of meeting the production plans that the computer was producing. Although the computer knew all about the parts that went into each product and the lead times involved in manufacturing or buying them, it had no idea of the work content involved and how this compared with the existing workload in the factory.

MRP II computer programs were supposed to plug these gaps. Many companies, however, have found that they are very complex to work with and take a long time to get right. In theory, it is a better approach than MRP I but a much more time-consuming and costly one to implement.

MRP II's fit with Just in Time is better than the fit with ROP but is *worse* than the fit with MRP I. Although it produces a production flow, it also forces management to follow a computer generated production schedule. Just in Time lays down its own—and very different—schedule. Just in Time and MRP II are not really compatible; companies must choose between them. And again, Just in Time helps companies improve quality and reduce costs – MRP II does not really address these issues at all.

OPT—OPTIMIZED PRODUCTION TECHNOLOGY

OPT recognizes that the problem MRP II is trying to address can be both improved and simplified. OPT recognizes that the only machines where capacity is really a problem are bottlenecks, where the production load exceeds available machine capacity. Nonbottlenecks, where the production load is less than the capacity, should not be a problem. The OPT solution, therefore, is to plan the bottlenecks in detail, relying on orders to flow naturally through the nonbottlenecks, Just in Time to keep the bottlenecks supplied with work.

This doesn't sound very different from what we've learned so far about Just in Time, and as OPT also helps companies to improve quality and reduce costs, the fit becomes even closer. Companies can quite happily have many of the ideas behind OPT and JIT working together in their factories.

SO WHICH IS BEST?

This is something that has perplexed many managers! The problem is, though, that it's the wrong question. The *right* question to ask is "Which approach is most appropriate?" because each approach is directed at a slightly different type of business.

Some businesses are very repetitive—the term mass production is sometimes used to describe them. Others are essentially jobbing shops, making either very small batches or one-offs. ROP, for example, is clearly an approach suited to reasonably repetitive businesses. There is no point holding stocks of parts for things you only make occasionally.

MRP I and MRP II, however, are slightly more versatile: while they are admittedly best suited to repetitive manufacture, they can in theory be used for one-offs if companies are prepared to set up each one-off in the computer every time.

Just in Time, though, can be applied to virtually all businesses. As we will see, it has techniques appropriate for both one-offs and mass production, in addition to a core of cost reduction and improved quality that is applicable to both.

EXERCISE

How repetitive is your business? Which of the above approaches best describes how your company works at the moment? What do you think are the key differences between the way your company currently operates and how it could work under Just in Time?

Chapter 2

Eliminating Waste

People sometimes think that Just in Time must be very complicated. This is totally untrue: the ideas behind it are very simple indeed. The most fundamental of these is the elimination of waste. Although eliminating waste doesn't sound very exciting, the definition of waste in Just in Time is much broader than simply wasted raw materials.

In Just in Time the definition of waste includes such wastes as the expenditure of unnecessary manpower effort and superfluous handling or needless waiting time. By constantly cutting out the unnecessary—the non-value adding activities, as they are sometimes termed—processes become simpler, faster, and less costly.

This chapter examines two of the four major areas of waste, manpower and machinery. The elimination of these two was Mr. Ohno's starting point.

Two important areas of waste:

◆ manpower

◆ machinery

MANPOWER

A common fallacy in business is to assume that people's effort is not being wasted as long as they are busy. This assumption is bound up in much of traditional management, and is one that can be very misleading. People are not necessarily being productive even though they are busy. It is possible for factories and warehouses to appear extremely efficient, with everyone diligently working away, yet still be full of wasted effort.

There are many reasons for this. Traditional payment and motivation systems, for example, are often based on motivating people to keep busy rather than on maximizing their productiveness. Another reason is that it is often easier for people to simply carry on with the old way of doing things rather than thinking of different and more efficient ways.

Just in Time helps with both of these by motivating people to work more efficiently and by putting increased emphasis on finding better ways of performing activities.

It is not difficult to spot wasted manpower effort once you start looking. There are many different sources of wasted effort. Here are four of the most common:

◆ unnecessary tasks

- unnecessary parts of tasks

- tasks that could be made quicker

- tasks that could be made easier

Unnecessary tasks

The key to locating unnecessary tasks is very simple. Ask the question "Why is this being done?" The longer it takes to find an answer, the more likely it is that the task is unnecessary!

As companies grow and change, activities are often put in place to solve particular day-to-day problems and needs. Part of every manager's job is to troubleshoot the day's problems by installing systems and procedures to avoid a recurrence. Few managers are quite so diligent at removing them once they are no longer needed! This is not really surprising—meeting challenges and solving problems is not only what managers are supposed to do but what their bosses measure them on. It's also fun. But unnecessary tasks arise through other causes as well.

When a procedure or part of a procedure is first designed, getting it to work often takes priority over simplification. Simplifying on the drawing board is difficult anyway; it is much easier to think things through after there is something concrete to look at.

New procedures or processes often have a belt-and-braces safety net built in too. Although with experience this safety net becomes unnecessary, people very rarely take the trouble to formally change things and design it out in the first place.

In spotting unnecessary activities, it is also sometimes useful to look at processes as a whole, rather than just at individual pieces. Sometimes activities are carried out in the erroneous belief that the next operation or process requires them. Five holes might be drilled in a component when only three are needed, for example, or information obtained and passed on that is not then used.

Parts might be unnecessarily counted or measured at the start of an operation when this had already been done as part of the previous operation—although the information was not passed on.

In the same vein, suppliers can incorporate initial operations on components as part of the manufacturing process of the raw component that they deliver. Holes in a component could be screw-threaded at the same time as they are drilled, for example. Materials handling activities are often found to be the simplest and most straightforward activities to eliminate. They are, by definition, non-value added—adding only cost and not value to whatever is being handled. Transportation, storage, and handling add nothing to the product—they simply contribute to costs and take up time.

Unnecessary parts of tasks

Sometimes, even though a task cannot be eliminated in its entirety, individual parts of it can. A form might still have to be filled in, for example, but the number of questions that are to be answered could be reduced.

In the same way as materials handling activities make good candidates for complete elimination, assembly and setup activities are often found to be capable of substantial simplification, when looked at in detail. The key, once again, is simply to ask questions that challenge the assumptions behind tasks. Instead of asking "Do we have to do it at all?" though, the question now becomes "Do we have to do *all* of it?" or "Can we simplify it at all?"

Components and setup alignments, for example, can be eliminated by the use of self-locating lugs so that parts simply pop into place. Repositioning of parts prior to assembly can be eliminated if they arrive already facing the correct way—a small saving, only seconds perhaps, but still an item of waste to be eliminated. From many such small savings, big improvements can subsequently be made.

Clerical and information handling activities are sometimes where the biggest savings can be made. Paperwork of all types has usually been designed with accuracy and completeness in mind rather than speed and simplicity.

Completing tasks more quickly

There is a natural limit to the extent to which tasks or parts of tasks can be eliminated. But there is virtually *no* limit to the extent to which they can be speeded up. As with eliminating tasks, there is a double bonus. Products flow through manufacturing and get to customers more quickly, and costs fall as well.

Assembly operations lend themselves very readily to improvements designed to speed them up. This can be done by using anything from power tools to pop-in fasteners that don't need screwing or tightening. Tell an engineer or an equipment supplier that you want something speeded up, and you won't be short of suggested solutions!

Product design is also important. Designers can be very talented at dreaming up products that work but are often less good at designing them so that they are easy to put together. Because designers and development engineers work in laboratories rather than on the shop floor, they can easily be out of touch with current production techniques—particularly if they've never studied them in the first place! Leading Just in Time companies employ two teams of designers—one to design the product and the other to redesign it for ease and speed of manufacture.

Repeatability is an important concept here too. The Just in Time company is interested in making operations *consistently* quicker, rather than *usually* quicker. Planning is easier if an assembly or setup consistently takes four minutes to perform rather than anything from ten minutes to an hour. People tend to work faster on repeatable and constant tasks too. A consistent pace of work can be set up and maintained as opposed to a stop-go, stop-go routine that slows people down.

Making tasks easier

There is often a lot of overlap between making tasks easier and making them quicker. A task made easier is usually a task made quicker. But it is also a task that is less prone to errors and one that is less tiring. Both clearly have their advantages.

A task that is mentally or physically difficult will cause people to make mistakes. Mistakes, in turn, require rectification, as well as the employment of inspectors to find them. Nor is rectification always possible—sometimes scrapping, or selling as seconds, is the only answer. Making tasks easier therefore helps to eliminate waste and also helps to improve quality.

Tasks that are mentally or physically difficult are also tiring. Not only do tired workers work more slowly, they also tend to make mistakes. Making tasks easier to perform is vital in eliminating waste.

Ten ways to make tasks easier:

1. Eliminate lifting wherever possible.

2. Use color and symbol coding to reduce decision making.

3. Provide checklists of optimum setup and operating conditions—machine speed, temperature and feed rates, and so on.

4. Use pop-in fasteners rather than threaded ones.

5. Use hand power tools as much as possible—and multiple tools (perhaps color coded to prevent confusion) to avoid changing bits or drills.

6. Use gravity—assemble downwards rather than upwards—even if this means turning the assembly upside down.

7. Use bolts that are as short as possible.

8. Keep workplaces tidy and uncluttered.

9. Have everything that is needed within arm's reach (rather than just a step away).

10. Replace all broken and worn hand tools that don't work the first time every time.

◆

MACHINERY

The idea of machines being wasted is initially rather a strange one to grasp. Unlike, say, a raw material or a period of time, a machine is in no sense used up. It is there at the start of an operation and still there at the end. How can it be wasted?

The trick is to regard the machine as one might a tool hired from an equipment shop. You have rented it; a notional meter is ticking away somewhere and management's task is to extract every possible ounce of value from the machine before it has to be returned.

Of course, many machines are manned while they are working or being worked on and so eliminating a manpower waste means eliminating a machine waste as well. In these instances, identifying and eliminating a particular waste is twice as beneficial.

Eliminating machinery waste is particularly valuable, however, because of the far greater cost of acquiring it in the first place. Most machines cost far more than people to rent. Also, as a machine's workload increases, a point is reached where there is fractionally more work than available capacity, even with, say, overtime and extra shifts. At this point, buying a whole extra machine is the only answer—it is impossible to buy half a machine in the same way that an extra half day's people can be obtained.

Three methods can be used to reduce machinery waste:

◆ reducing operation times

◆ reducing setup times

◆ eliminating overproduction

Reducing operation times

An operation time, or cycle time, is not as fixed as is commonly supposed. People are often surprised at the extent to which Just in Time companies are able to reduce their machines' operation time. Surely, they argue, the biggest factor—the machines' speed—is set by its manufacturer. This argument is usually wrong, because it overlooks the fact that the real operation time is the floor to floor time: the elapsed time between picking a part or component up and putting it back down, having completed the operation on it. For much of this time, the machine's set speed is irrelevant. Analyzing typical operations in detail, it soon becomes apparent that for a large percentage of a machine's supposed operation time, it is not really operating at all—it is waiting.

Picking the part up, for example, takes time. Positioning it on the machine ready for the intended operation takes time. Fastening it in place and unfastening it after the operation takes time, and so does putting it back on the floor and getting ready to pick up the next one.

Just in Time companies achieve remarkable reductions in operation time by largely dispensing with all these activities. Suppose, for example, that instead of having to pick parts up from floor level and put them back down, the parts were simply picked up from and put back down onto a simple roller track or conveyor? And instead of needing to be carefully positioned and then fastened in place, they were put into a self-locating and quick-clamping jig? Although none of this sounds very dramatic at all, the results might look something like this:

A Typical Drilling Operation

	Number of Seconds to Complete	
	Before	*After*
Pick up	5	1
Locate	4	2
Fasten	10	3
Drill	5	5
Unfasten	8	3
Put down	5	1
Total (seconds)	37	15

Not bad! With a little expenditure on basic handling equipment, the operation time has been reduced by 60 percent. Not only have we speeded the job along—a batch of 1,000 now takes a whole six hours less to complete—but we've created extra capacity on the machine that can be used to make products for other customers. Note that the actual drilling time has remained unchanged. All that has happened is that by eliminating waste in the handling process, we have enabled the machine to spend more time doing what it was designed to do—drill. This sort of waste elimination is at the very heart of Just in Time.

Reducing setup times

In the case of setup time reduction, Japanese companies such as Toyota have been able to achieve enormous improvements. Machine presses, for example, that used to take hours to set up, now take the same number of minutes—a sixty-fold reduction!

The ways in which setups can be speeded up are virtually limitless, ranging from the application of simple common sense to complex engineering. The key is to recognize the distinction between internal setup time—when the machine physically has to be stopped—and external setup time, when it need not be.

Much time can be saved through making sure that all the tools and equipment needed for the changeover are at hand before the machine is stopped. Putting peg-boards of tools by each machine avoids the need for walking off to hunt down that elusive spanner halfway through. *Use a scale graphic internal vs external*

Although a lot can be achieved by making sure that as much as possible of the setup is done in external time, there comes a point where nothing further can be done until the machine stops. The effort should now be focused on doing everything possible to speed things up, for production has now come to a stop—as opposed to external setup time, when it hadn't.

Many of the techniques that we have already looked at are equally applicable here—eliminating lifting, for example, by the use of roller tracking to slide tooling dies in and out. But there are other tricks, too: cartridge-type single position connectors, multiple-connection plugs and twist-and-lock fasteners instead of bolts. All save time, minimize the need to make subsequent fine-tuning adjustments, and help to prevent errors.

Of equal value is effort invested in making setups achievable by one person whenever possible. Not only will the actual operation be slightly faster—unless well rehearsed, people tend to get in each other's way—but it also cuts down on wasted waiting time. As with trying to arrange meetings, it's harder to get three people together on time than two. The more people involved, the greater the waiting time.

Eliminating overproduction

The final area of machine waste is overproduction. At first sight, this too is deceptive. Obviously, producing unnecessary parts is a waste, so why would anyone do it? What can be saved by cutting out what shouldn't happen anyway? The answer is surprising, for overproduction is indeed far more widespread than it should be.

The trick is to recognize that it usually tends to be cleverly disguised as something else! Most operations have a built-in overproduction allowance, for example. This is a loss or scrap factor that is built in to the batch quantity to allow the operator a few components for fine-tuning the settings on the machine or for a few scrapped and rejected parts in the course of the batch.

So, if 100 parts are actually required, for example, with a 5 percent overproduction allowance the actual number of parts going through an operation would be 105.

Clearly, if the need for the extra five parts could somehow be eliminated, machine capacity could be freed—not just at *this* machine, but at all the other machines ahead of it in the operation sequence that also had to produce the extras. This multiplication factor is why eliminating overproduction is such a powerful Just in Time technique.

Good graphic

What's more, overproduction actually compounds up. If each of six machines in an operation sequence need a 5 percent overproduction allowance, then the machine at the start of the sequence needs to produce a whole 34 percent more than the real requirement! For example, with a final batch size of 100, the machine at the start of the chain needs to make 134, from which the next machine produces 128 and so on, to yield 100 after the sixth machine. Efforts made to reduce overproduction pay big dividends.

Overproduction also occurs in the way that materials are used. Raw materials such as rolls of material or sheets of steel from which individual components are cut or punched are often processed only in units of whole rolls or sheets. This means that if an order calls for only half a sheet of material to be used, the full roll will be used anyway. The extra parts go into stock until another order comes along—provided of course that one does come along and that the parts aren't lost or damaged by then.

While the machine is processing these unwanted parts—remember, there is no customer's order or requirement for them—it cannot be producing parts for which there are orders that could be generating sales income and profit. Despite this, traditional companies often regard it as wasteful to only process half a sheet of material.

EXERCISE

What forms of overproduction are there in your business? What purposes are they supposed to serve? Make a note of them: in the next chapter we look at how to harness skills within the organization in order to eliminate them.

Chapter 3

Continuous
Improvement
Teams

Clearly, most companies have plenty of areas where they can reduce waste. But how can one person manage to do it all—and how will he know when to stop?

The answer, of course, is that one person *can't* do it all, and there isn't a single point at which all waste is eliminated so that one can sit back and relax. Instead, waste is identified and eliminated by teams of people, often called continuous improvement teams, who bring together the variety of skills necessary to do the job. And, as the name implies, continuous improvement teams strive to eliminate waste continuously. Their task never ends!

Although manpower and machinery wastes are generally regarded as the most significant, wastes of time and material are also important. Continuous improvement teams often find these wastes while they are looking at those in manpower and machinery. Indeed, working on these wastes first is sometimes a good way for a team to cut its teeth—they are usually easier to resolve, and early success helps the team to work together.

In this chapter, we are going to look at how to establish continuous improvement teams. There are five aspects to creating a successful continuous improvement team.

◆ understanding the strength of a team

◆ selecting the team

◆ motivating the team

◆ training the team

◆ leading the team

UNDERSTANDING THE STRENGTH OF A TEAM

Teams are much more than simply collections of individuals. For a start, they are far more powerful and capable than individuals. When teams work together they can build on each other's contributions, creating a solution that is better than the sum of the individual contributions.

Academic research with problem solving exercises, for example, has shown that almost two-thirds of teams will perform better than the best individual team member on their own and that virtually all teams will perform better than average team members on their own.

Teams also have another attribute: because the solutions that they produce are joint solutions, they are far more likely to come up with approaches and decisions that are generally acceptable to everyone concerned. As a result, changes suggested by a team are usually more likely to be successful than changes suggested by an individual.

SELECTING THE TEAM

Managers sometimes look at the sort of results achieved by Just in Time companies and despair of finding people in their own companies who can achieve the same results. Not so! All it takes is the sort of people likely to be found in *any* company or organization.

Despite this, there are certain types of people that it's helpful to have on board, especially when just starting out with Just in Time. Equally, there are one or two types of people to avoid, especially at the start of an implementation. (Both types are described later in this section.) Although, when first establishing the idea of continuous improvement teams in your organization, it's best to stack the odds in your favor by picking certain types, in general, most of the people employed by your company are likely to make competent team members.

THE POSITIVE APPROACH

The chief attribute to look for is one that is sometimes described as a can-do type of personality. Seek out individuals whose initial approach to problems and difficulties is *positive*: people who look for ways of solving problems and improving things, rather than for reasons why change is either impossible, difficult, or inappropriate at the moment.

A word of warning: dynamic, action-oriented individuals, while having forceful personalities, are sometimes overly dominant. Be careful not to have a collection of would-be dictators who will quarrel with each other. That way, nothing is achieved except disharmony.

It doesn't matter too much if the people you have in mind are initially not interested in Just in Time or don't believe that it will work. Once they have seen what can be done, some truly amazing conversions will be observed. In fact, as with other aspects of life, some of the best enthusiasts start out as disbelievers. Given an open mind, though, the results will soon speak for themselves and you will have gained not just a convert but a disciple eager to preach the message to others.

An important personality type to avoid is the blocker. These are the people who block progress and change, either actively or passively.

The active blocker

Active blockers have various ways of standing in the way of change. These range from a shake of the head and a gloomy prediction that "it'll never work here," to a grudging acceptance that although it might be a good idea in theory, it is bound to fail because of a number of small operational difficulties.

The passive blocker

Passive blockers can be more subtle. Their approach is sometimes described as the soggy sponge technique: they never stand in the way, but never actually get around to implementing anything. Proposed changes always seem to need a little more consideration; there are always a few more people that need to be consulted, and so on.

Like active blockers, passive blockers are bad news: both can be overridden by the sheer momentum of a successful implementation once it gets going but can be fatal in the early days.

Skills and backgounds

The other important point to bear in mind when selecting people is the breadth of skills and backgrounds that they can bring to the team. This too is more important in the early stages than later on. To begin with the team needs to be self-sufficient. Later on, it will be confident enough to co-opt people to help as it needs them.

The exact blend of skills and functions depends upon each team's focal area: some teams will be pointed towards production, others to shipping, some to purchasing, and so on. The important thing is that the team should be broadly based.

The composition of a team looking at all apsects of production, for example, might be

◆ production foreman.

◆ two production operators or lead persons.

◆ production control person.

◆ purchasing person.

◆ manufacturing engineer.

◆ design engineer.

◆ you.

Each person is there to contribute to the total picture on behalf of their function. So the design engineer, for example, is there to offer views on how a product's design could be altered in order to make it simpler to manufacture. More than this, though, they are there to actually perform—wherever possible—those functions themselves. For example, if the team identifies improvements that could be achieved by a design change, rather than have the change go to the back of the design department's backlog for three months, the can-do design engineer takes it on herself.

MOTIVATING THE TEAM

Naturally, you can't just assemble a team and tell them to get on with it. Depending on the general level of morale within the organization—and also within the team itself—some element of *motivation* will be required.

In some organizations, particularly one with a more autocratic culture, the establishment of a team will itself be seen as an unusual departure and members may be reluctant to join in. In other organizations, the establishment of the team may be viewed as simply the latest management flavor of the month, resulting in members feeling reluctant to commit themselves too deeply to the project.

Management style

The manner of the motivation itself depends heavily on a manager's style. If your style is *persuasive*, for example, you will want to win the team over with logical arguments about the need for change, the potential for improvement, and so on. Alternatively, if your style is more *assertive*, you may simply say: "This is where we're going: follow me."

Whatever your management style—and there are a great many—some form of rational communication of the team's objectives is required. The argument put forward here should act as a helpful model, particularly if coupled with appropriate statistics where available.

THE ARGUMENT FOR BUILDING THE JUST IN TIME TEAM:

◆ The organization needs to improve its performance.

◆ Many of the areas that need fixing fall within the scope of Just in Time.

◆ Just in Time has worked for a great many companies.

◆ There's no reason why it won't work here.

◆ The organization's problems have more than one root cause—it's not just production, or purchasing, or design, but a combination of all of them.

◆ Rather than one big speculative jump we're going to take some small more manageable steps instead.

◆ You are therefore creating a multifunction, multiskilled team to take some of these small steps.

◆

TRAINING THE TEAM

Even when motivated, the team will still need to know what is expected of it and how to go about achieving it. Essentially, the team needs training in

◆ Just in Time.

◆ problem solving and planning techniques.

◆ teamwork and communication skills.

Just in Time training

The extent of the training that will be required depends very much on the individuals involved in each team, and tends to differ from team to team. Clearly, all teams need to understand Just in Time itself, and so a short appreciation course—possibly based on *Understanding Just in Time*—will be required, perhaps supplemented by further appropriate reading material.

Naturally, different people will have different needs at this point; engineers might want to read more about short setup time techniques, while production people might be more interested in the machine linking and production control techniques that we will cover in Chapter 6.

Problem solving and planning techniques

Some knowledge of problem solving techniques will almost certainly be required, and here the fishbone diagrams that we will look at in Chapter 4 will be useful.

The main objective is to get people to think about the causes of what is going wrong—the *real* causes, rather than what they are told by other people. Some basic planning tools are also helpful. There is no point in a team rushing off to tackle a dozen different things all at once and bogging down without finishing any of them.

Teamwork and communication skills

Consideration also needs to be given to the extent to which teamwork and communication skills need to be improved. Again, this is something that calls for judgment based on the precise composition of the team—and not a little tact! Particularly in a production environment, team members may have had little experience in sitting around a table and formulating ideas, let alone in communicating them to others via a flip chart or blackboard. The danger is that the less experienced team members are likely to be less confident in voicing their opinions and less able to present them in a coherent way that makes their point. As a result, the team's view of an issue could be dominated not by the people who are most knowledgeable about it, but by those who are most confident about saying something.

In general, while training is important, it needs to be kept in perspective. Unless real problems develop, training—apart from the actual Just in Time training—is unlikely to be a make or break issue. It's usually better to be moving ahead and *doing* something instead of just training.

LEADING THE TEAM
Leadership, however, *is* a make or break issue. Over the years, perhaps even more has been written about leadership than has been written about motivation. With Just in Time continuous improvement teams, the right leadership is vital.

Three areas in particular are important:

◆ giving the team the space to develop solutions

◆ barrier-busting

◆ giving the team the freedom to fail

Giving the team the space to develop solutions

When leading a Just in Time team, it is important to understand that you can specify a goal or a target—"we will halve this operation time," for example—or you can specify a time—"action by this time next week!" You can't, however, do both simultaneously.

Traditionally, managers are more time orientated than results orientated. Watch out for this. With Just in Time, it is better to set an aggressive, hard-to-meet goal than it is an aggressive, hard-to-meet timetable.

The nature of Just in Time, as well as the experiences of those companies that have achieved the biggest improvements through its introduction, means that it is best simply to set the team goal and give them the time and space to develop their own ways of achieving it.

It is also vitally important to avoid imposing your solutions on the team—not only will that destroy their initiative and lower their morale but it will also tend to elicit an inferior solution. The great advantage of the team, remember, is its diversity, from which can emerge something that is greater than the sum of the parts. It may be unpalatable, but swallow hard, sit back, and wait!

Barrier-busting

Perhaps your biggest challenge will be breaking through the various obstacles that the team encounters or has had put in its way by outside blockers. This is where you have to come to the rescue. The team can't progress any further on their own: it's up to you to add your weight to knock down the obstacle (or steer a course around it) and allow progress to be resumed.

The approach to doing this depends on one's individual style: some managers will prefer the use of reasoned dialogue with those of their peers who comprise the barriers; others will seek a route around it or simply over it. The important thing is to get the team to the other side: failure detracts both from your credibility and the team's motivation. Fail often enough and the whole implementation will stall.

Giving the team the freedom to fail

It is also vital that you as leader understand that the team's approaches to a solution can sometimes be trial and error. Things simply don't work the first time every time, and changes may sometimes backfire.

The answer is not to berate your team—especially in the face of extensive criticism—but to carry on moving forwards. "OK, we've learned from that. Now how can we do it differently?" is a far better response than a dressing down or admonition.

EMPOWERMENT OF PEOPLE

The empowerment of employees is now being recognized as a key factor in the competitiveness and ultimate long-term success of American companies. Robert Rosen, a psychologist who writes about leadership and healthy organizations, promotes the idea that all people in an organization are intellectual assets who should be mobilized by enlightened leadership to harness their competence, creativity, and commitment. There is no question that companies who have taken this path have become superior in their field and renowned for their high quality operations and products. (Toyota, Harley Davidson, Honda, Saturn, Xerox, and Boeing).

Thomas Kochan and Paul Osterman, professors in management at MIT, in their book *The Mutual Gains Enterprise*, stress the critical importance of a mutual management, labor, and government approach to the empowerment of the U.S. workforce as being necessary to sustain the competitive gains that U.S companies have recently achieved in the world market. They feel that the restructuring and downsizing that some companies have undertaken in the 1990s is demoralizing to the workforce, and is often carried out to produce cash in the short term—cash used to increase the salaries and perks of top management—with no long-term improvement in the company's prospects.

Chapter 4

Total Quality

Having assembled, trained, and motivated your team, you are now ready to lead it into action. A good place to start is with the third of the four wastes—wasted material.

Actually, it's slightly broader than that. We are going to look at the role of quality in the Just in Time organization—and of course quality impacts on far more than just efficiency of material usage! Poor quality results in not just scrap and rejects, but wasted effort in rework or remaking, higher than necessary inspection levels, and likely customer dissatisfaction. It also results in wasted marketing and sales effort too—sales representatives who can't sell to customers because of the poor quality of the products the last time that they bought some, and so on.

For all these reasons, improving quality is a good place to commence. It is also a good place to start for two other reasons. Tackling quality and material waste allows the team to cut their teeth on problems that are likely to be both simpler and faster to fix—a good confidence builder.

Even more important, good quality is vital if Just in Time is to work. With traditional approaches to manufacturing, where products took, say, four to five weeks to make, there was often enough slack time to remake them, if it turned out that they failed final inspection. With manufacturing times cut to a day or so with Just in Time, there isn't the scope to do this.

In this chapter, we are going to look at:

◆ the causes of poor quality.

◆ operator involvement.

◆ identifying quality problems and sources of material waste.

modify graphic

THE CAUSES OF POOR QUALITY

Understanding what causes poor quality in the first place is a big part of understanding how to prevent it. Poor quality is a variation from a standard: the standard being, ideally, perfection. But what causes the variation?

Variation in a typical component might be due to

◆ variation in the machinery speed.

◆ variation in tooling dimensions.

Include in Mod 1

◆ variation in setup.

◆ variation in the material of the component itself.

◆ variation in the operator's technique.

◆ variation caused by contamination and damage.

◆ variation caused by miscellaneous other factors.

Some of these are random influences or nearly so. Variation in setup might be regarded as such, as might variation caused by contamination or damage. Others, however, are not. Machine speed and tooling dimensions don't move erratically from component to component—they might vary from the standard but do so by gradually drifting over a period of time.

Acceptable quality level

The traditional approach to quality control tended to regard all variation as random. Old style quality control was in fact more accurately termed inspection, where sampling exercises determined whether batches of components were acceptable or not. Hence the name given to the cutoff level of the various samples: AQL or Acceptable Quality Level. An AQL level of 95 percent, for example, meant that provided that only 5 percent of the parts were rejects, the batch was acceptable. There were two principal problems with this approach.

First, management's target or objective became simply to produce components or products at the AQL level. Provided that they didn't have to throw away more than, say, 5 percent, then they had done their job. As we know, that argument clashes with the Just in Time approach of eliminating *all* waste (not just up to a 5 percent level) and aiming for continuous improvement. Why be satisfied with 95 percent—why not 96 percent or 97 percent or even 100 percent?

Second, because of the randomness assumption, the traditional approach didn't assess quality until after the batch was complete. Make all the parts; *then* check them against the standard. But where variation from the standard gradually drifts in, with a drill bit that gradually wears thinner, for example, this can actually be tested for and determined before the batch is finished—in fact, before any defective parts are produced, if trapped early enough.

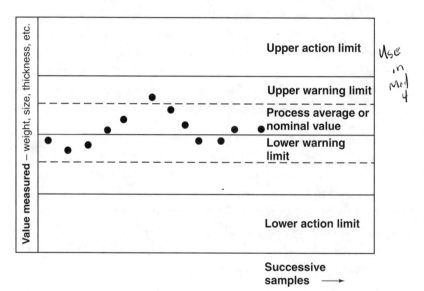

Figure 1
Illustration of Statistical Process Control Diagram

Statistical process control

Statistical Process Control (SPC)—now common in the West, but popular even earlier in Japan—is a widely used technique for trapping these drift problems. Parts are periodically measured—by the individual operator, usually—and the appropriate dimension plotted on a piece of graph paper by the machine. As drift occurs, the line on the graph paper clearly moves. Well before the drift reaches the critical point, action can be taken to correct the problem and bring the component back into nominal tolerance.

OPERATOR INVOLVEMENT

The Just in Time approach to quality extends the SPC idea in two vital areas. First, it extends the measurement of variation from standard into those supposedly random areas such as contamination damage and setup problems. Second, it moves the concept of quality away from being something that is inspected-in, achieved by testing a batch after it has been manufactured, to something that is built-in, by making the operator feel responsible for the quality of the products or components that he is manufacturing.

This is an important point. The basis of traditional manufacturing is that workers are regarded as having little pride in their work, that their tasks are de-skilled, and that those who perform them are supposed to be motivated solely by factors such as piecework and bonus schemes. In reality, though, operators actually want to produce quality products and components. No one actually likes producing defects; everybody wants to improve the quality of what they are doing if they possibly can.

Just in Time harnesses and fosters this basic natural instinct. The first thing to do is, of course, to communicate to the operators the fact that they are responsible for the quality of their output. This needs to be done intelligently: it is no use simply putting up lots of posters saying "quality is very important" and stopping there!

Quality charts

The message can be put across in a number of ways. Many companies make use of people's innate competitive nature and put charts next to individual operator's machines or workplaces that graph or log their quality performance. This makes quality a personal issue, as opposed to being someone else's problem. In addition to harnessing people's competitive spirit, it also makes use of another social pressure: the fear of being seen failing. The charts are a very public indicator of just how well people are performing.

For this reason, caution is needed when implementing them. It is no use if management merely insists on putting charts by people's machines; there first has to be a groundswell of opinion that it would be a good idea. In any case, operators should fill in the charts themselves, as part of the process of being responsible for their quality. This needs an atmosphere of cooperation rather than one of confrontation.

As we will see in Chapter 6, another way of bringing home the quality message is where operators can clearly see the consequences of poor quality. This is at its starkest when parts are simply of no use, having to be rejected and so holding up the production process until they can be replaced or reworked.

In traditional manufacturing, where the next operation on a batch of parts may well take place some days (or weeks) later and in perhaps another part of the factory altogether, this linkage is usually lost.

Putting processes together

By moving to a system of putting production processes much closer to each other, however, the implications of poor quality become readily apparent. If a component has holes drilled in it in one operation, and the next operation, carried out immediately afterwards, is to then bolt that component onto another component, it quickly becomes apparent if the holes are too small or out of alignment.

The real advantage of this technique is obvious: Moving operators and machines closer together not only saves time (as we will see in Chapter 6), but eliminates waste as well; the wasted material and the wasted effort in wrongly drilled holes can be eliminated. Instead of having made a whole batch of faulty parts, the operator will have only made a handful or so before the fault is identified in the first operation. The problem can be rectified immediately without any more faulty parts being produced.

Even those few parts could have been eliminated, however, by making operators inspect their own output. There is often no need for complicated SPC schemes—in many cases simple go/no go devices are just as effective. In this case, one such device could be a sample of the component that the part was to be bolted onto, with bolts protruding from the matching bolt holes. By simply marrying the two parts, the operator could easily check that they would actually fit together and that the holes drilled were of the right size.

IDENTIFYING QUALITY PROBLEMS AND SOURCES OF MATERIAL WASTE

Naturally a continuous improvement team will find it straightforward to cut its teeth on relatively simple single improvements like this. But improvements like these don't really harness the full power of the team.

Where the team comes into its own is in identifying opportunities to eliminate waste, rather than in developing solutions. For finding the precise cause of a problem is a very large part of the battle: subsequently coming up with a cure is often quite straightforward.

Thinking back to the idea of quality problems as unwanted variations, the task of the team is to probe deeper and deeper into these variations. Some of them, like tooling or setup variations, might be easier to deal with than, say, variations in the precise composition of the raw material. The causes of some problems, of course, may be initially completely unknown, being regarded at the moment as simply random variations. The task of the team is to look at variations in quality and eliminate them, one by one.

Fishbone diagrams

The Just in Time process for doing this involves what is usually known as a fishbone, or cause and effect diagram. It is also known as an Ishikawa diagram, after its Japanese inventor, Dr. Kaoru Ishikawa.

Fishbone diagrams are useful tools for dealing with a wide range of problems—their application need not be limited solely to quality problems or sources of material waste. Another common application, for example, is in identifying the causes of lost time in the production process, so that delivery times can become more responsive.

Defining the problem

The diagram starts with the statement of a problem—for example, cracks in a component. Whereas traditionally, some cracked components would have been regarded as inevitable natural waste, a continuous improvement team will use a fishbone diagram to identify and eliminate all the sources of variation that lead to the cracks. This is where the team plays a vital role. It is unlikely that any one individual will have all the knowledge and insights that are required; in a team, though, they do.

First, the team lists the main areas in which variation can occur—this is sometimes called the coarse fishbone. It looks like this:

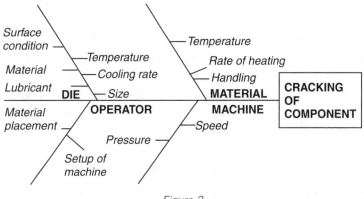

Figure 2
The Coarse Fishbone Diagram

Although there are only four bones in this particular diagram, there may often be many more, depending on the problem in question. Some diagrams may have ten or more. Generally, the less specific the problem, the more bones the coarse cause and effect diagram will have.

Next, the team brainstorms the various aspects of each of these main bones, listing them as subsidiary bones of their own to produce a fine fishbone diagram. The diagram might now look like this:

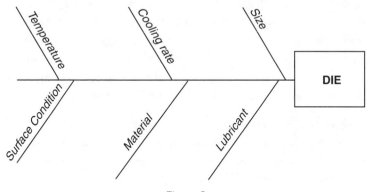

Figure 3
The Fine Fishbone Diagram

Of course, each of these identified factors may have very fine bones of their own. Take die lubricants, for example:

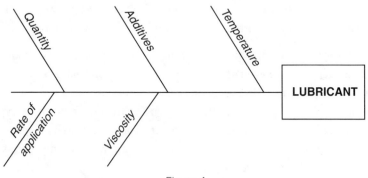

Figure 4
The Die Lubricant Diagram

Having identified each of the various causal factors making up the fishbone—or at least as many as they can—the team examines each factor's influence on the problem. Sometimes this can be relatively straightforward to determine: the broad direction of the influence might be already known but perhaps no one has put in place any action to prevent it from causing problems. Problems may be caused by the temperature of a quenching oil, for example. It might be known that at too low a temperature the likelihood of parts distorting increases. This information is useless, however, if the operator doesn't have a temperature gauge so that it can be checked—or is unaware of the precise temperature limit in the first place.

Sometimes, however, the effects of the various influences are genuinely unknown. They have been listed on the fishbone diagram because they seem to have a bearing on the problem. Are they a *real* influence or a phantom one? And if there is a real cause and effect mechanism, exactly how does it work?

At this stage, experiments are necessary, testing the various alternatives to see how each of them has a bearing on the problem. Some experiments the team will be able to conduct for itself. Others will perhaps need specialist technical investigation—something that might require the team leader's authority to set in motion.

A good tip is to rank the factors in a rough order of magnitude, then tackle the ones at the top first. Factors don't have an equal impact. The top half dozen or so can sometimes cause the vast majority of the quality variations. Focusing the efforts of the team on these will have a significant impact. Those that aren't considered so important can be put aside for later, until the team's resources are less stretched.

Chapter 5

Just in Time
Purchasing

So far, we have looked at eliminating waste and striving for continuous improvement within our organization. But few companies work in isolation. Most businesses need to buy components and materials to make the products that they sell and many of them find that the proportion of total cost represented by purchased items is generally increasing.

This is even truer for companies and organizations that buy virtually all that they sell, such as distributors. As a result, the role of purchasing is becoming more important and the Just in Time company has to work closely with its suppliers if the hoped for objectives are to be achieved.

Some companies—especially those with high levels of bought materials—actually start implementing Just in Time by looking first at the purchasing function. Many companies find that applying Just in Time purchasing techniques is the fastest way to achieving some Just in Time benefits. Whereas continuous improvement and waste elimination programs on the shop floor can take time to come to fruition, the simpler Just in Time purchasing techniques can often deliver benefits immediately.

This chapter explores these benefits and how to achieve them and then build on them. The ideas behind Just in Time purchasing are very similar to those that we have looked at already. A common theme is *waste elimination*—not only the familiar wastes of manpower, machinery, and material; but also wasted time and wasted space. The basic elements of what companies need to do are very simple.

◆ Work with suppliers.

◆ Reduce the supplier base.

◆ Improve supplier performance.

THE BENEFITS OF WORKING WITH SUPPLIERS

The traditional approach to purchasing was to regard suppliers as adversaries. Purchasing managers saw their roles as being to exact the keenest possible prices from their suppliers, threatening them with loss of business if they did not cooperate. The same goods were often obtained from two or more sources, and orders placed with the cheapest. Prices were driven down—and performance driven up—by threat. The relationship between customer and supplier was an arm's length one: "This is what we want—do it or else."

In no sense can this traditional approach be described as working with suppliers to achieve a joint objective! It is, in fact, very wasteful. Time and effort is expended in haggling and arguing with suppliers about quality problems and late deliveries, for example, and still more time is wasted with delivery times that are long and inflexible. Equally, the traditional way of securing prices—quantity discounts—simply generates wasted space through having warehouses full of raw materials that aren't needed yet.

In contrast, the Just in Time way of working is based on having longer term relationships with suppliers and treating them as partners rather than adversaries. The two companies then cooperate to achieve an enhanced performance, with benefits for both parties.

The particular business benefits to be aimed for differ from Just in Time company to Just in Time company, but there is usually a common core:

TYPICAL BENEFITS:

◆ lower stock levels

◆ smaller, more frequent deliveries

◆ shorter lead times

◆ simpler systems

◆ better quality

◆ lower costs

Single sourcing

The starting point is single sourcing. Many companies have traditionally had two or more suppliers for every part or product that they bought. In Just in Time purchasing, companies work out which supplier is best for them (on whatever criteria) and give them all their business, rather than just part of it. Prior to Just in Time, this was unusual and was considered risky.

In fact, it's actually *less* risky. For the intended result of this single sourcing is mutual dependency: each party should be so important to the other that cooperation and commitment become essential. When more and more of a supplier's business is sold to a single customer, it is vital not to let that customer down. Equally, when a company gets all of its particular widgets from a particular supplier, that supplier essentially becomes an extension of the customer's own factory. *This* is how the benefits are achieved. Naturally, consistent and reliable planning and ordering systems are vital, because there's no one else to fall back on.

Safety stocks

When customers and suppliers work together, many of the safety stocks that companies hold can be dispensed with. Suppliers are able to work on a reliable schedule, and no longer need to hold high levels of stock to cushion themselves against erratic customer demands. Customers know that their suppliers will deliver on time, so they can also reduce their safety stocks of raw materials.

The traditional approach to purchasing is to place large orders with suppliers in order to get discounts on the price. Just in Time customers don't want deliveries in large lumps. The Just in Time approach is to get suppliers to deliver a flow of parts, ideally matched to the actual production rate, rather than in delivery quantities that may equal many months' worth. Again, this is easier to organize when a single company supplies a range of parts rather than just a few.

Not surprisingly, with less stock in the system, faster response times are possible; innovations can be introduced more quickly and market trends followed more closely.

Management time

There can be savings in management time, too. Many companies find that they can simplify their ordering and control systems. For example, an engineering company might decide to buy all its fasteners from a local supplier who offered a merchandising service. The supplier visits the company once a week to replenish its stock. Whatever has been used in the previous week gets replaced; the supplier simply invoices for the fasteners that he's left at the company and returns the next week.

The company finds that there's no wasteful consumption of management time in deciding how many of each fastener to order, no purchase orders to be raised, and no goods inwards functions to be performed. The company explains that it doesn't place purchase orders for the electricity or gas that it uses, so why should it for the fasteners that it needs? It can concentrate on what it does best—being an engineering company—and it's the job of the supplier to make sure that the fasteners are on the shelves when they're needed.

The quality impact

Quality improves, too. Not only do Just in Time companies consciously deal with only the highest quality suppliers, but they also try to develop those suppliers still further. Moving suppliers away from the traditional inspection approach towards techniques such as Statistical Process Control improves quality as well as reduces costs. Encouraging them to adopt Just in Time waste elimination programs of their own helps even more.

REDUCING THE SUPPLIER BASE

Having fewer suppliers is clearly the key. But how should companies go about achieving it? The starting point is to recognize that the best supplier is not necessarily the cheapest. Saving a few cents on the price but then incurring additional costs in poor quality, unreliable deliveries, high inventories, and let-down customers is not good business. Deciding which suppliers to keep and which to lose is not a simple exercise; few companies have all the information that is needed to do this. One of the leftovers from the traditional approach to purchasing is that companies usually have far more information about their suppliers' prices than about their reliability or the integrity of their quality and planning systems.

To reduce the supplier base properly, careful analysis is necessary. The starting point is usually a data gathering exercise, to fill in some of the gaps in the picture. When that is complete, companies usually visit their major potential suppliers to assess them face to face.

Checklists of criteria against which to assess suppliers varies from business to business. Businesses outside of the engineering industry, for example, are not going to be too interested in their supplier's in-house engineering facilities. For a company in the automotive industry though, engineering capability is often vital. The importance attached to the different factors also differs from company to company: delivery frequency, for example, is more important for some businesses than for others.

Initially, the key thing is to simply draw up a checklist of the factors that are important to a business and to attach rough provisional ratings to them. Refinement can come later.

ASSESSING SUPPLIERS: A TYPICAL CHECKLIST

◆ How much do we spend with them?

◆ How local are they?

◆ How frequently will they deliver?

◆ How reliably will they deliver?

◆ What are their prices like?

◆ What is their quality like?

◆ How good are their planning systems?

◆ How good is their engineering?

◆ Are they innovative?

◆ How significant will we be to them?

◆ How serious are they about us?

Only rarely will a clear leader emerge on all points. Quite commonly, it is necessary to make trade-offs, settling for slightly poorer performance on one or two characteristics in order to attain better performance on others. At this point, some of the longer term aspects come into play. Can the supplier actually be developed, so becoming the best on *all* scores?

Size has a bearing on this. It is easier to educate a supplier into your way of doing things if you are an important customer—yet another reason for single sourcing, and also one for selecting smaller suppliers, with whom it is naturally easier to be in a stronger position.

Another point in favor of smaller suppliers is that smaller companies often have more flexible systems, so there's no "main office won't let us do it that way" interference. But size is a double-edged sword, for it's a strategic issue too. Smaller companies aren't necessarily the leaders in their industry in terms of innovation and product development. They are also regarded as being riskier in terms of viability.

IMPROVING SUPPLIER PERFORMANCE

By having fewer suppliers, significant improvements can be made in each supplier's performance. The closer relationships enable the same sort of waste elimination and improvement programs that operate inside the company to work outside it as well. The impetus for achieving performance improvements comes from two sources:

◆ supplier assessment

◆ continuous improvement teams

Supplier assessment

Supplier assessment—sometimes known as vendor assessment—was operated by some of the West's better companies before the advent of Just in Time. Many of the actual techniques are the same: the key difference lies in the purpose to which the assessment data is put.

Traditional companies see supplier assessment as another part of the big stick approach that they favor. If we collect information about how badly our suppliers perform, they say, we can then threaten them with losing the business to a competitor or demand a price reduction.

In contrast, Just in Time companies collect some of the same information but instead use it to drive continuous improvement programs, perhaps jointly with the supplier herself. The assumption is that the supplier is already the best or most suitable one available and that there is little point attempting to reestablish another relationship with an alternative supplier, who may only be marginally better. Instead, Just in Time companies tend to look at how the present supplier can be helped to make substantial improvements.

Supplier assessment is still worthwhile even if there is nothing actually unacceptable about suppliers' performances. Rather like runners recording their times so that there's something to aim at, having recorded performance data offers the opportunity to try and pare that little bit off next time. Without it, companies are aiming in the dark.

The sort of performance data that Just in Time companies record varies according to the type of business involved, but there's usually a common core:

Supplier challenge slide

◆ elapsed time between order and delivery

◆ quoted lead time

◆ number of defects per million parts delivered

◆ response time in the event of any problems

◆ frequency of delivery (weekly, daily, or hourly)

◆ amount of inventory in the pipeline

The last of these—the inventory pipeline—is particularly interesting. Companies sometimes fool themselves into thinking that they're achieving Just in Time if they are getting the delivery performance that they want but with the supplier having to achieve it by holding stock. This of course is wrong. As we learned earlier in the book, the objective is to get stock out of the system entirely, rather than move it from one company's premises to another. If your supplier isn't truly manufacturing Just in Time himself, then he won't be achieving the elimination of waste that brings the quality, delivery, and cost improvements that you, the customer, deserve. Having less inventory is only part of the picture.

SUPPLIERS, ISO 9000, AND TOTAL QUALITY MANAGEMENT

The advent of the European Economic Community, with its elimination of trade barriers and tariffs among a powerful group of industrial countries, along with the keen competition from quality oriented Japanese companies, led the European companies to adopt total quality management and other methods to improve their own operations and products. However, companies in different countries operated to design, production, test, inspection, service, and quality standards that were in many cases unique to that country.

The Europeans decided, therefore, to create a set of quality, or more accurately, customer expectation standards that would be universally used by all the companies operating within the European Community (formerly European Economic Community), and ultimately the rest of the world. This standard is known as International Organization for Standardization (ISO) 9000.

The most comprehensive of the five standards within ISO 9000 is ISO 9001, which is the major standard, and it confirms process conformance from the initial development of a product through production, test, installation, and servicing. ISO 9002 covers standards only for procurement, production, installation, and servicing. ISO 9003 addresses standards for final inspection and testing only.

To be certified for any of these standards, a company must set up internal process standards targeted to consistently achieve a high level of quality and customer satisfaction, suitable for its own business, comply with those standards, and be audited for the existence of the procedures and standards and compliance with them. ISO 9001 is by far the most meaningful of the three standards since it covers the entire product development process.

Some 40,000 European companies have now been certified to the ISO 9000 standard, and because U.S. companies do extensive business with European companies, some 2,000 to 3,000 U.S. companies have now been certified to the ISO 9000 standard.

The universal acceptance of the ISO 9000 standards in the United States and throughout the industrial world makes this a significant quality standard. The ISO standard emphasizes a quality program and defines twenty total quality management areas that are subject to audit during a company ISO 9000 certification process:

1. management responsibility

2. quality system

3. contract review

4. design control

5. document and data control

6. purchasing

7. control of customer supplied product

8. product identification and traceability

9. process control

10. inspection and testing

11. control of inspection, measuring, and test equipment

12. inspection and test status

13. control of nonconforming product

14. corrective and preventive action

15. handling, storage, packaging, preservation, and delivery

16. control of quality records

17. internal quality audits

18. training

19. servicing

20. statistical techniques

The ISO 9000 process is a powerful tool in the achievement of the sustainable high quality levels necessary to successful implementation of Just in Time inventory control methods.

Continuous improvement teams

The other impetus for change comes from your continuous improvement teams. As they identify opportunities for waste elimination and improvement within the production process, they are likely to want to feed ideas through to suppliers. This is obviously one reason why there should be a representative from purchasing on the team. But the representative need not be the sole conduit of communication. With a good relationship between customer and supplier, the supplier will want to send her people in to talk to the team directly. The team, in turn, may want to visit the supplier.

The areas picked for examination obviously differ from company to company. Typical improvement issues include the following.

◆ Can we organize daily deliveries of the components needed for today's assembly program?

◆ Can we arrange for a truck to tour a group of our suppliers, collecting what we need from each of them in order to reduce costs?

◆ Can we ask our suppliers to use minimal packaging—why pay for their people to pack components only for ours to unpack them?

◆ Now that quality is predictable, can we have deliveries straight to the production line, rather than through receiving and inspection?

◆ Can the parts arrive ready to use, in returnable packaging?

HOW TQM IS CHANGING U.S. COMPANIES

In the 1980s Ford Motor Company management invited Dr. Deming to assist them to become more competitive and has now developed within their entire organization a total quality management system that has enabled them to continuously improve their products and adopt a Just in Time procurement system. Their costs have been significantly reduced, and they began to aggressively take significant market share from General Motors and the Japanese auto companies.

Chrysler Corporation adopted full implementation of total quality management, employee empowerment, and Just in Time in the development of their highly praised and extremely successful redesigned minivan introduced in the 1996 model year. They continue to make considerable progress towards full company implementation of total quality management.

At General Motors, the Saturn Division has since its inception adopted total quality management, employee empowerment, and Just in Time techniques to the full and using these techniques introduced a top quality automobile to the marketplace. Saturn has consistently scored in the top four in vehicle quality and customer satisfaction surveys conducted in the United States.

Boeing Aircraft Company has, over the last ten years, made steady progress with their adoption of total quality management and continuous improvement techniques in their aircraft building operations and in their now much smaller supplier base. Inventories have been reduced and component rejections and rework are now at an all time low.

Chapter 6

Putting It All Together

So far we have looked mostly at improving and eliminating waste from individual parts of the production process. We have seen how costs can be reduced and quality improved by the efforts of individuals and teams working towards the goal of continuous improvement.

A production process, though, is a long chain of events and operations. While we have looked at ensuring that each link in the chain is as reliable—and as short—as possible, we have yet to look at the best ways of joining them.

It is only when these links are put together that the full potential of Just in Time is reached. This chapter details two ways of creating the links:

◆ Close coupling of machinery

◆ Connections with *kanbans*

CLOSE COUPLING OF MACHINERY

We saw in Chapter 2 how production could be speeded up by cutting out as many as possible of the material handling operations or by making them easier and quicker to perform even if they couldn't be eliminated entirely. As we saw, quite basic changes, such as simply avoiding the need to pick up and put down parts or components between operations, can make a big difference.

Just in Time aims to eliminate all these wastes of time and effort and brings sharp reductions in the total cycle time, thus speeding delivery to the customer.

Weaknesses of the traditional approach

Unfortunately, with the traditional approach to manufacturing, there is a limit to the extent to which this is possible. Consider for example the typical factory layout. It tends to be departmentally or functionally laid out. All the lathes are together in one department, for example, and all the drilling machines in another and so on.

There are several consequences of this. The first is the sheer distance involved in a typical part's manufacture, which even in a small factory can soon mount up to hundreds of feet. Not only does this waste effort, energy, and manpower—it has to be transported, perhaps by a person driving a forklift truck—but it takes *time*.

Time is also wasted due to bins or boxes of parts never being transported to the next operation as soon as they're ready—they always have to sit and wait until the transport is arranged. Even more tellingly, and far more wasteful of time, is the fact that individual parts don't move at all—only batches of parts! The first part that is put into the bin has to wait until the last one is put in before it is moved, which may take hours or even days.

EXERCISE

Take a typical part in your factory. Plot its progress from operation to operation, and department to department. How far does it travel? How long does it take to transport it? How much time does the first part to be processed spend waiting for the last part to be finished?

Another problem with this type of layout is that it is bound to involve the operatives themselves in wasteful handling activity—

certainly in picking the next part to be processed out of the jumbled ones in the bin and also (most probably) in picking up and putting down activities as well. There has to be a better way.

Cells and U-shaped lines

There is. Just in Time avoids all these wastes by coupling machines together into groups—not groups of similar machines but groups to make similar parts or products. These groups, or cells, are sometimes called product-oriented rather than process-oriented cells.

In the ideal world, each cell would make a single product. This is wasteful, though, unless there is enough demand for the product to justify keeping space and machinery tied up to make it. However, the concept works just as well with families of similar parts instead.

The way that it works is straightforward. A number of parts or components are identified that all go through (perhaps with the odd exception) the same machines in (more or less) the same sequence. The closer the match, the better, although there is room for a little flexibility.

Next, the machines are relocated, and put side by side. They are not, however, put in a straight line: they should be in either a circle, or, even better, a U-shaped line. The importance of the U shape is that it enables operatives to reach more easily—and therefore operate—more than one machine. In traditional manufacturing, multiple machine manning is rare unless machine cycle times are deemed long enough to warrant it—usually at least two or three minutes. During this two or three minutes though, the operative is idle—and therefore being wasted. U-shaped lines help avoid this.

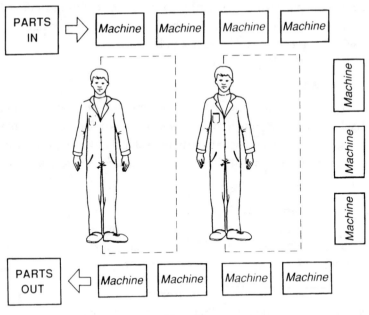

Figure 5
Schematic diagram of U-shaped line

Handling equipment

Once positioned, simple tracking or roller belts are put in place between the machines, to move individual parts from one operation to the other. The logic here is to avoid wasting human effort by having operatives do it, even though the actual physical effort may be minimal. If possible, simple ejection devices are then installed on the machines to kick out the component onto the tracking or belt as each machine finishes it. Again, wherever practicable, the objective is to save as much human effort as possible, in order to free up time for multiple machine operation. In some cases it is possible to have devices that not only eject the parts but also put them in place, so that all the operator has to do is press the start button; however, this is usually harder to achieve. Particularly at first, it is best to aim simply to have the operators put the part in place on each machine and press the start button.

It obviously all takes effort to achieve this. Having done so, though, the benefits are enormous. In sheer productivity, for example, the U-shaped line can repay the costs of creating it very quickly. Where previously you may have had eight machines, each employing an operator who spent a great deal of each machine's cycle time simply waiting for operations to be complete, you now have two operators who are working productively all the time.

Work in progress

Inventory levels will also plunge. Previously, you might have had a binful of 500 parts at each machine. Eight machines, 500 parts at each—that's 4,000 parts of inventory. Although the accountants might have called it work in progress, clearly most of it wasn't any such thing, It was "work in waiting," sitting in the bottom of the bin until the rest of the batch was finished!

In contrast, the U-shaped line has a single part *genuinely* in process on each machine, and one lined up on the track by each machine. With time, that one part that is lined up can also be eliminated. From an inventory level of 4,000 to one of 16 is quite a sizable reduction and clearly one worth investing some effort to achieve. But there's more.

A lot of wasted time has also been eliminated. Processing work with bins full of 500 parts takes time. Exactly how much time varies enormously from factory to factory. Days or possibly weeks—it all depends on the slickness of the scheduling system and the distances between machines. With the U-shaped cell, however, the time taken before parts start flowing off the line is down to a matter of minutes. The time taken to complete the whole batch will be little more than 500 times the sum of the cycle times—a matter of hours rather than days or weeks.

EXERCISE

> *Think about the part or component that you looked at on page 76. How much time is taken to manufacture a batch at present? What might it be with a U-shaped line?*

A further benefit is also that quality will improve. As we saw in Chapter 3, where parts move directly from one machine to the other, faults can be trapped right away and corrected. If a problem with a part is found part way through, the line can be stopped, the fault identified and fixed without incurring either wasted time or significant numbers of parts needing scrapping or rework—perhaps only half a dozen or so at most.

Clearly, the close coupling of machines brings substantial benefits, making the effort of relocating machinery into U-shaped lines very worthwhile. One commonly found difficulty, though, is that companies find that they haven't quite got enough machines to form the number of lines that they need. Although they should in theory be roughly in balance, there is inevitably a need for an extra drill or lathe or some such item. It is important not to fall into the trap of thinking that this means that Just in Time is somehow less productive—it can't be with the sort of performance improvements we saw earlier in this chapter. It is simply a line balancing problem. There are two answers. The first is to buy another machine—with performance improvements like this, the capital justification should not be too difficult! But as the machines in question are usually quite minor ancillary ones, many Just in Time companies find that they actually have what they need already, perhaps in mothballs, or lying unused in another part of the factory. Although old, they are perfectly serviceable: they might be a lot slower than their modern equivalent, but quite adequate given the performance potential of the U-shaped line as a whole. If there isn't a suitable machine lying around, then secondhand ones can be purchased quite cheaply. Buying a new one should be a last resort!

CONNECTIONS WITH *KANBANS*

Sometimes, though, direct physical linkages aren't possible. Large expensive pieces of plant that feed production into—or receive it from—a number of separate product lines are a case in point. A paint line, for example, would not normally be suitable for building into a U-shaped line. Other examples include processes like acid cleaning and casting, where the environment of an acid shop or foundry is such that the traditional departmental layout is best retained.

Outside suppliers are another example of where physical linkages just aren't possible either. Yet the benefits of linkage are undeniable: so an alternative method has to be employed.

Many Just in Time companies use the concept of *kanbans* to achieve linkage and product flow in situations where direct physical coupling won't work. The term *kanban* simply means ticket or sign in Japanese, and the concept was pioneered in Toyota by our old friend, Mr. Ohno.

Kanbans take advantage of the short cycle times that Just in Time can achieve—for example, by the use of U-shaped lines and through the elimination of waste—to show that it is possible to produce parts virtually on demand. This is in complete contrast with the more traditional approaches to manufacturing scheduling that we looked at in Chapter 1.

ROP systems, for example, don't even attempt to produce a production schedule for products or parts based on the demand for them: they simply aim to keep increasing stock so that there's something available (either finished products or the parts that go into them) when demand occurs.

While MRP systems attempt to do this, they too need enormous amounts of computer power (even more than reorder point systems) and also tend to produce schedules that don't reflect real capacity and workload limitations within the factory.

People sometimes get confused about *kanbans*, imagining them to be much more complicated than they actually are. Using examples, let's look at the two most common types of *kanbans*:

◆ *kanban* squares

◆ *kanban* cards

Kanban squares

Imagine a Just in Time factory where you have a group of machines performing operations on a number of different parts that come from another group of machines next to it. Management doesn't want to directly couple the machines into U-shaped cells, but still wants the benefits of linkage. But how will the link "know" how many of which component to feed through from one group to the other? Whereas traditional manufacturing thinks in terms of solving this by internal production orders or computer systems, many Just in Time factories employ a much simpler solution. They paint *kanban* squares on the factory floor between the two groups of machines, each square the size of one bin of components. The square has the component's part number on it and may even be color-coded with the same color as the tooling and setup instructions in order to facilitate rapid changeover.

The operation of the *kanban* squares is simplicity itself. As bins of parts are used, the painted squares that they stand on become empty. This—and only this—is the instruction to make another bin of the same parts in order to once again cover the square. The system is simplicity itself, with no paperwork or management instructions needed to initiate manufacture. If the square is empty, it is filled. If not, the operator either produces something else for which there *is* demand—in other words, an empty *kanban* square—or finds another task altogether. Whatever happens, she doesn't make something in advance of its *kanban* square becoming empty, just to keep busy: that would destroy the linkage, and with it the benefits of flow.

Kanban cards

Now, let's extend the *kanban* concept to the entire production chain. Imagine that an order for a finished product comes in to your Just in Time factory from a customer who wants, say, a left-handed widget. There is one on the shelves—just as there might also be in a factory using MRP or reorder point systems—and so it gets shipped. This, however, exhausts the stock: there aren't any more. This too is exactly as it might be with reorder point or MRP systems. Now, though, the procedure in the Just in Time factory differs.

The warehouse foreman actually initiates the manufacture of more widgets himself instead of simply waiting for them to arrive, having been ordered by the system. He takes the *kanban* card from the warehouse shelf where the widgets are stored and hands it to the assembly foreman. The *kanban* card appears quite simple: plastic covered so that it can be used again and again (the elimination of waste, remember!), it says something very straightforward such as "make another batch of 500 left-handed widgets." There will probably also be the widgets' product or part number printed on the card and also a simple drawing of it—to help unambiguously define exactly what is wanted and avoid any confusion with right-handed widgets.

The assembly foreman is very pleased to get the *kanban* card: his workers have three or four batches of something else to make first and then they will be able to start making the widgets. *Kanban* cards, just like *kanban* squares, act as *instructions to make*. There will be no paperwork or official works orders to accompany the instruction: the recyclable *kanban* is itself all that is required for him to have the authority to make some widgets. He therefore starts checking the components that go into the widget. Some of them he has; others he doesn't. For the components that he needs, he simply locates their *kanbans* and hands these to the foremen responsible for making them.

These foremen in turn then begin to assess the situation: starting the manufacture of those components where raw material is available and issuing raw material *kanbans* to stores or suppliers for those where it isn't. From the issuing of the initial *kanban* by the warehouse foreman, not only has manufacture of the required widget and its components been initiated—together with purchases of raw materials and externally purchased parts—but the whole process has been very simply organized. No computers, no schedules. And parts have only been produced when required, in the quantity required and in the order required—something traditional systems usually can't manage even with computers and schedules.

Chapter 7

Implementation

We have now looked at most of the actual mechanisms of Just in Time: the elimination of waste, the setting up of continuous improvement teams, and the establishment of flow through physical linkage and *kanbans*.

In this chapter we look at two final pieces of the jigsaw: identifying where to start and overcoming likely obstacles.

◆ establishing a pilot project

◆ obstacles to progress

ESTABLISHING A PILOT PROJECT

Why do you need to have a pilot project at all? Strictly speaking, of course, you don't. But most Just in Time implementations go more smoothly if the concepts are tried out first in a pilot area of the factory. Thereafter, implementation of the rest can proceed at whatever pace is felt to be comfortable.

It's important to note that the idea behind having a pilot project trial isn't to make sure that Just in Time works. That has been proved in company after company around the world. Instead, the wisdom of establishing Just in Time in a pilot area first is mostly to do with factors inside the implementing company itself.

A more manageable task

Getting a pilot project working is easier than trying to convert the whole company all at once. This is a sensible idea for many reasons. It will probably be people's first experience of Just in Time, for example: going for something smaller and more manageable is clearly a good idea. Undoubtedly, various problems will be encountered en route—we will look at some of these later in the chapter— and these are usually easier to resolve if they affect only a pilot area. Management's time, too, is a limited resource. If full commitment is to be given—and this is vital—then this too is more easily obtained with a pilot area, which can be used to some extent as an experiment, rather than with a company-wide implementation where the resistance to change, even at the top, can be much greater.

It is usually much quicker to get Just in Time working in a small pilot area first, and faster results are a useful fillip, encouraging people and motivating them to press on. It's too easy for a long project to simply run out of steam and bog down! Establishing U-shaped cells, continuous improvement teams, and *kanban* systems takes time; thus, the smaller the area to be converted, the faster progress can be made.

Not all aspects of Just in Time are equally applicable to all companies. The relative importance of Just in Time purchasing or setup time reduction, for example, varies from company to company. Choices need to be made, too. Should you use SPC? Do you need *kanban* squares or *kanban* cards, or both? And should this U-shaped line be laid out this way or that way?

In general, the approaches that are settled on in the pilot area will usually be those that are most appropriate for the other areas, too. The pilot area thus acts as a working model. This not only makes the implementation of successive areas easier—"just copy what we've already got working"—but helps in estimating likely lead times and reduced space requirements.

A demonstrator

A demonstrator is also useful for convincing all the doubters and blockers who don't believe Just in Time can work. It is difficult to disbelieve the evidence of your own eyes, and the pilot area's improvements will be plain to see. If necessary, a board or notice listing former and Just in Time lead times, quality levels, floor area, and productivity will help to drive the message home.

Selecting the pilot area

Ideally, the pilot area should be as self-contained as possible. The intention should be to try and model the full implementation, so it should ideally take in genuine raw materials at one end (as opposed to part-processed materials or components from elsewhere in the factory) and produce finished goods for shipment to customers at the other.

It isn't always possible to achieve this. If necessary, compromise on the production of finished goods: it's more important to have control over the start of the process than it is to produce finished products for customers. The idea is to try and achieve isolation from the rest of the factory. The pilot project needs to be seen as successful and so shouldn't run the risk of being jeopardized by late deliveries of materials from elsewhere in the factory.

Size is another important factor. Ideally, the pilot area should be small enough to be a manageable experiment, but big enough to be significant. Doubters and blockers need to be convinced by the results, which they are unlikely to be by an implementation covering only 1 or 2 percent of the factory's output. Somewhere between 10 and 15 percent is a much better size to aim for.

Try also to select a pilot area that is as ordinary as possible. Avoid exotic processes, unique technologies, or anything else that the doubters and blockers could use to say, "Ah, well, it won't work here."

The implementation team

The implementation will of course need a team to carry it out—it's too big a task for an individual. Many of these issues that we looked at in Chapter 3 with respect to continuous improvement teams apply here, so there's no point repeating them. The implementation team is usually simply a more senior form of a continuous improvement team and tends to work in much the same way. Only the objective of the team is different: achieving a complete implementation rather than identifying improvements.

It's quite possible for people to be on the project implementation team as well as on a continuous improvement team—this can, in fact, be a useful channel of communication. Leading by example also comes in here: a prime task of the implementation team is to get continuous improvement teams established in the first place—something that may be easier if implementation team members volunteer to become involved in them to help start the ball rolling.

The larger the company, the more complicated the reporting structure can get. Some companies appoint steering committees and project directors in addition to the manager and his colleagues on the implementation team. The trick to working with these additional committees or directors is to keep them involved. The absentee landlord approach is dangerous. The steering committee or project director not only won't have a good enough grasp of what is going on to be able to offer help when required, but may also hinder progress by soaking up the team's time with demands for special briefings and reports.

Planning and launching the pilot project

Obviously, a pilot project is a departure into the unknown. This means that having a good idea of where you want to go is vital, but ironically, also means that designing a route map is going to be harder. So, just as with a real journey, have a good idea of the destination but allow plenty of time to get there. Giving yourself plenty of time the first time around avoids the risk of panic measures creeping in in an attempt to catch up, or people becoming discouraged through imagining that the implementation is running out of steam. The project plan needs to have some slack built into it to cater for time lost going down blind alleys and for exploring different alternatives. Second time around, you can go a lost faster because you know the route.

Communication is also vital. Things are going to be changing— changing faster and more dramatically than people usually expect. To avoid alienating or worrying people, keep them informed. There's nothing more likely to generate resistance than the feeling that you're being "kept outside" or that your job is being endangered. Good communications not only forestall difficulties, but can actually speed the implementation. People are much more likely to pitch in and help things along when they know what's going on.

A key job of the team is to communicate, not just what it's doing, but what it intends to do. Despite the extra workload, it's best to try and formalize the communication, rather than relying on word of mouth on an "as and when" basis. Successful ad hoc communication is more of an art form than many managers realize. It's easy to mislead unintentionally or leave something out.

Some companies, especially larger ones, publish a periodic Just in Time newsletter to make sure that the message gets across. Other companies find that a Just in Time bulletin board, located where everybody can see it, works well.

Bulletin board items:

◆ photographs of the team

◆ copy of project plan

◆ explanation of what Just in Time is all about

◆ message of commitment from the top (vital!)

◆ photograph of pilot area

◆ photographs (or examples) of pilot area's products

◆ initial lead times, scrap levels, and so on—and the new targets

◆ progress-to-date information

Deciding the emphasis

Virtually everything that we have looked at in this book applies to any company wanting to implement Just in Time. The relative emphasis that should be placed on the various aspects will differ greatly though. Think back to the various alternative approaches that we looked at in Chapter 1—such as ROP, MRP II, and so on— and the level of repeatability that each of them is more appropriate for (mass production vs. job shops, for example). Think also of the weaknesses of your present business: where is it falling down? What do customers, and suppliers, complain about most?

Clearly, the nature of your business—the level of repeatability vs. one-offs—and the type of systems already in place will determine your priorities. For some businesses, setup time reduction will be more important than purchasing; other businesses might instead target *kanban* cards or the resolution of quality problems.

These issues won't always be easy to decide, and may generate quite heated debates. Everybody will have their own pet hates to exorcise and the debates may take some time to resolve. Don't be tempted to rush this process. A clear (and sensible) project plan makes an enormous difference to eventual success.

A word of warning: don't be tempted to duck the difficult issues and simply implement those aspects of Just in Time that nobody objects to. As we will see in the next and final section, Just in Time runs *counter* to traditional management's vested interests. If nobody objects to it, it probably isn't worth doing!

OBSTACLES TO PROGRESS

Sometimes Just in Time projects stall, having hit a barrier. We have already looked at some of the barriers to starting Just in Time; we will now look at two important impediments that may be encountered once the project is underway. These are

◆ traditional performance measures.

◆ output-based incentive schemes.

Traditional performance measures

Traditional ways of measuring performance don't sit very well with Just in Time. In some companies, both management and systems are flexible enough not to allow this to become a problem. In other companies, where the traditional measures are looked upon as being written on tablets of stone, they can become a significant barrier.

The problem is that traditional measures look at only part of the picture. Clearly, a Just in Time factory is far more productive than a traditional one. Traditional measures don't always show this, however. With Just in Time, companies manufacture in smaller, more responsive batch sizes and incur more setups as a consequence. Unfortunately, traditional productivity measures all report smaller batch sizes and more setups as being a deterioration in efficiency. This is because they measure output over input and the input half of the equation always includes an allowance for setting up, or its equivalent.

So while the Just in Time team is trying to reduce batch sizes and set up more frequently in order to be more responsive to customer demands, the company's traditional measurement systems are actually penalizing people for doing this!

Old-fashioned simple output measures also cause problems. Although factory managers have been measured on their output since time immemorial, the Just in Time company isn't aiming solely at sheer output, irrespective of whether it's destined for the stock shelves or customers' orders. The Just in Time company aims to *meet customer's demands*—a very different matter from just churning out as much as possible whether due, overdue, or early. By offering a premium service—what customers want, when they want it—Just in Time companies can make more profit than their traditional competitors who are still focused simply on churning goods out of the factory door.

Neither problem is easily solved. It's essentially an either/or decision and there isn't really and halfway house. You can't have Just in Time *and* old-fashioned measurement systems.

Output-based incentive schemes

The same is true of traditional payment schemes that motivate operatives to hit output targets. This again is only part of the picture. Operatives soon locate the trade-off point where they can make the greatest output at the lowest acceptable quality level.

Obviously, this runs counter to the idea of continuous improvement and the elimination of waste. The Just in Time philosophy is to reduce total costs, not just labor costs. What's the point of saving money on salaries only to spend more money on scrap, rework, inspectors, and annoyed customers?

The closer to pure piecework that a bonus system is, the more likely it is that it will cause problems. Again, it's an either/or issue: either the scheme is taken out—or at least neutralized—or Just in Time implementation will stall.

Surprisingly, taking such schemes out causes few difficulties, at least with operatives, who are usually among the first to point out how counterproductive they actually are. It's often managers and supervisors who worry most. This might be because bonus schemes are often used as a substitute for good management, rather than as an adjunct to it. And as we have seen in this book, Just in Time is built on good management—on empowering people to perform at their best, continually striving to deliver that little bit more.

INDEX